This book belongs to:

Made in the USA
Monee, IL
05 December 2020

50995569R00066